# WAVE AND RIPPLE
## DESIGN BOOK

Wave and Ripple Design Book
Copyright © 2019 by Satoshi Nakamoto

Published by
LEAF STORM PRESS
Post Office Box 4670
Santa Fe, New Mexico 87502
U.S.A.
leafstormpress.com

To inquire about bulk discounts for special events, promotions and premiums
please contact the publisher by email at leafstormpress@gmail.com.

First edition.
Printed in the U.S.A.
Book Design by LSP Graphics
SNPI-4999999837

Library of Congress Control Number:2019943290
Cataloguing In Publication data is available.

ISBN 978-1-9456520-3-5

# WAVE AND RIPPLE
## DESIGN BOOK

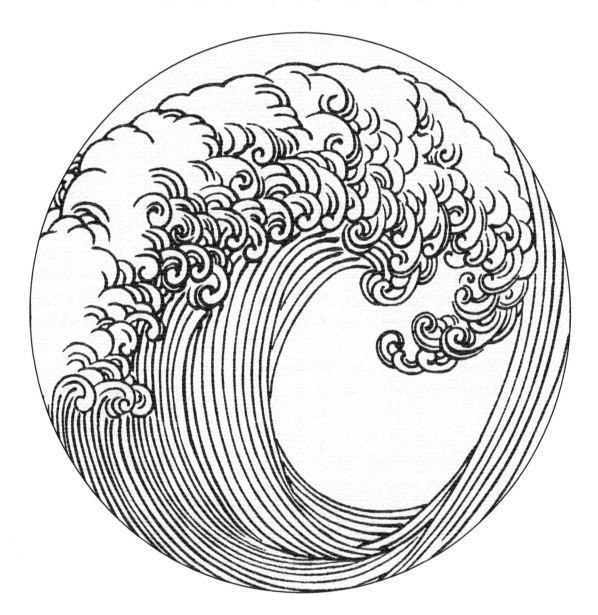

# SATOSHI NAKAMOTO

**Illustrations based on designs by Japanese artist Mori Yuzan**

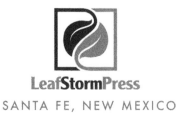

**LeafStormPress**

SANTA FE, NEW MEXICO

*Run from what is comfortable.*
*Forget safety.*
*Live where you fear to live.*
*Destroy your reputation.*
*Be notorius.*
**–RUMI**

# INTRODUCTION

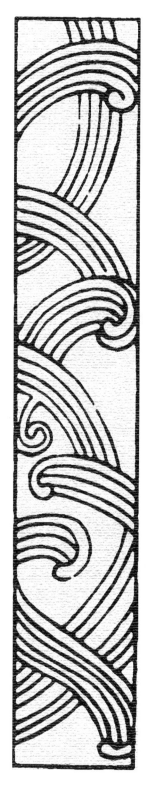

When you think about Japanese art, chances are it is Hokusai's iconic paintings, like his "Great Wave Off Kanagawa," that first come to mind. Lesser known, but equally masterful, are the black ink drawings of Mori Yuzan from the Meiji period. Done in the *Nihonga* style, Yuzan's designs were frequently used by craftsmen to adorn their wares. Although put to a more utilitarian purpose, Yuzan's drawings clearly rise to the level of fine art. His mastery of the line, the simple way he captures the dynamic quality of waves and ripples and conveys their energy and visual complexity is remarkable.

Yuzan's drawings and sketches are now readily accessible online, and images of his work are occasionally posted on social media, where they are quickly consumed by our content hungry, screen addicted culture. Mr. Nakamoto decided that Mori Yuzan's work deserved a more fitting tribute. One that would provide a more contemplative experience, and one that emulated the process of the craftsmen who used his designs. Except that, instead of adorning other objects, the art itself would be the sole focus.

After Mr. Nakamoto selected his favorite drawings, the original versions were restored, as needed. Then, adjustments and enhancements were made, with care taken to preserve those essential imperfections of Yuzan's work that make it so uniquely his. The final step in the process was creating a small edition of new illustrations for Mr. Nakamoto's private collection and reproducing them in this volume.

On behalf of Satoshi Nakamoto and myself, I hope you receive as much pleasure from this book as we did in the making of it.

–J.L.B.

*Postscript: I am an independent curator of rare books and manuscripts who was retained to assist Mr. Nakamoto in publishing this book of Yuzan's designs. In my field, it's a given that clients expect a high degree of discretion. In Mr. Nakamoto's case, however, preserving his anonymity required an array of precautionary measures straight out of a spy novel. That's why this introduction is not signed with my name, only with initials–initials that are not my own.*

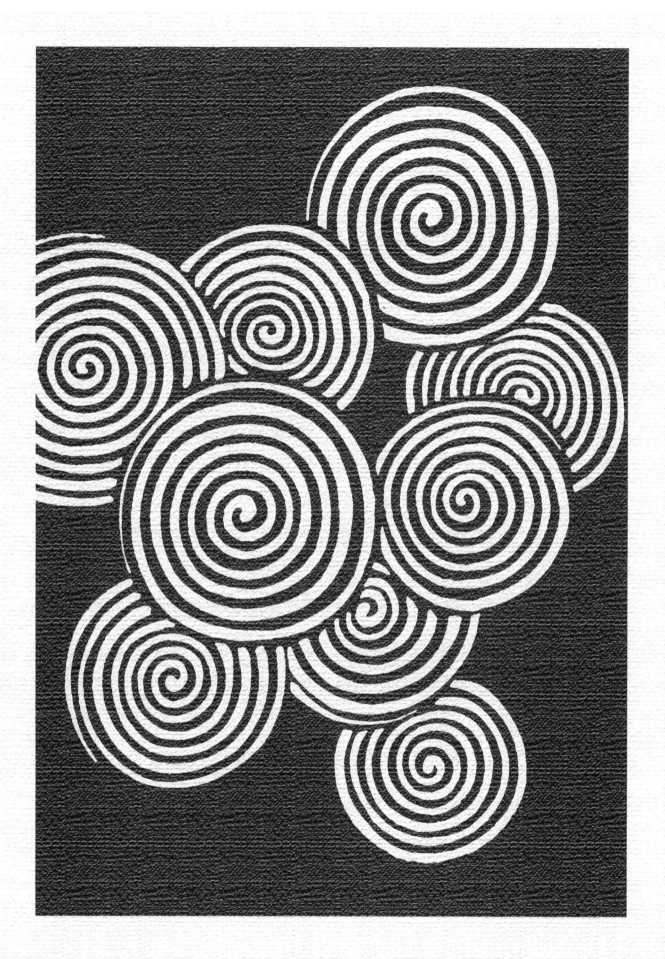

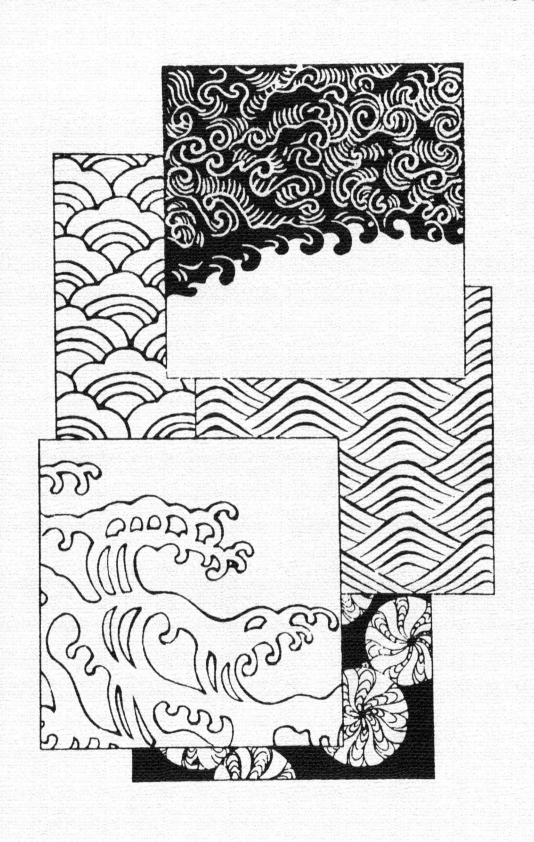

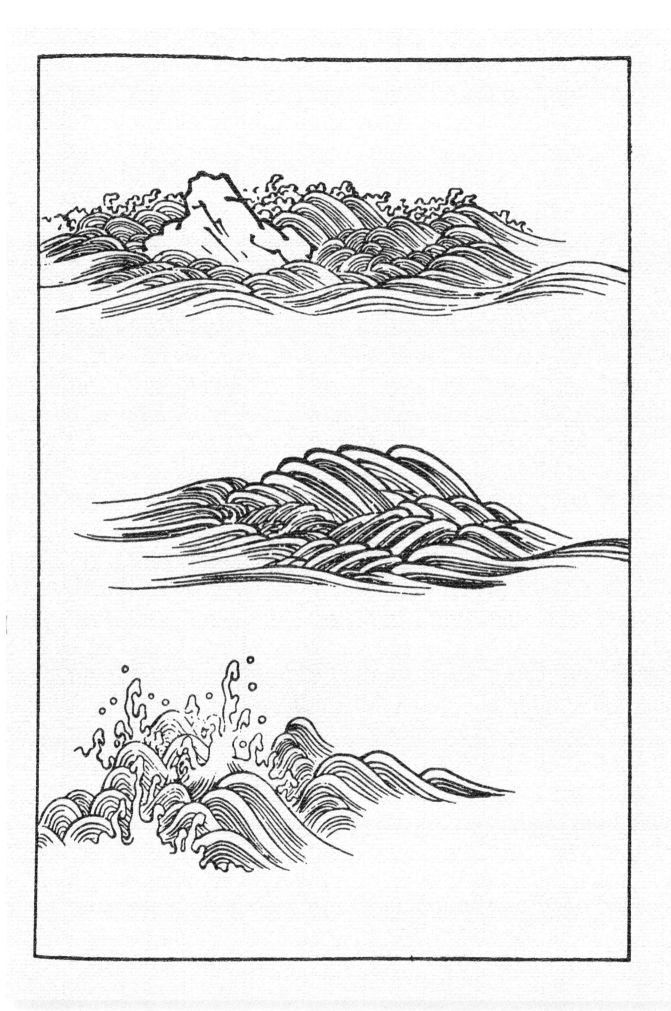

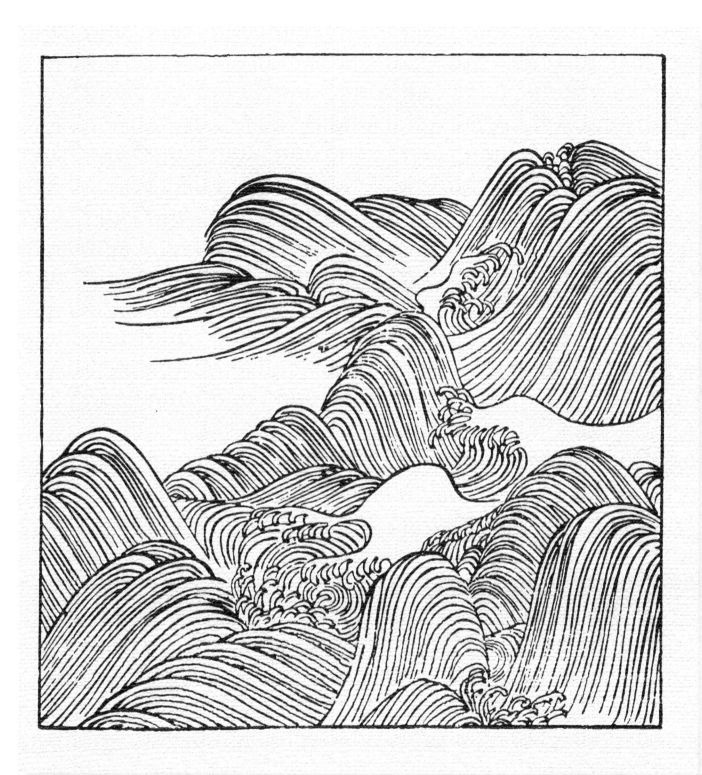

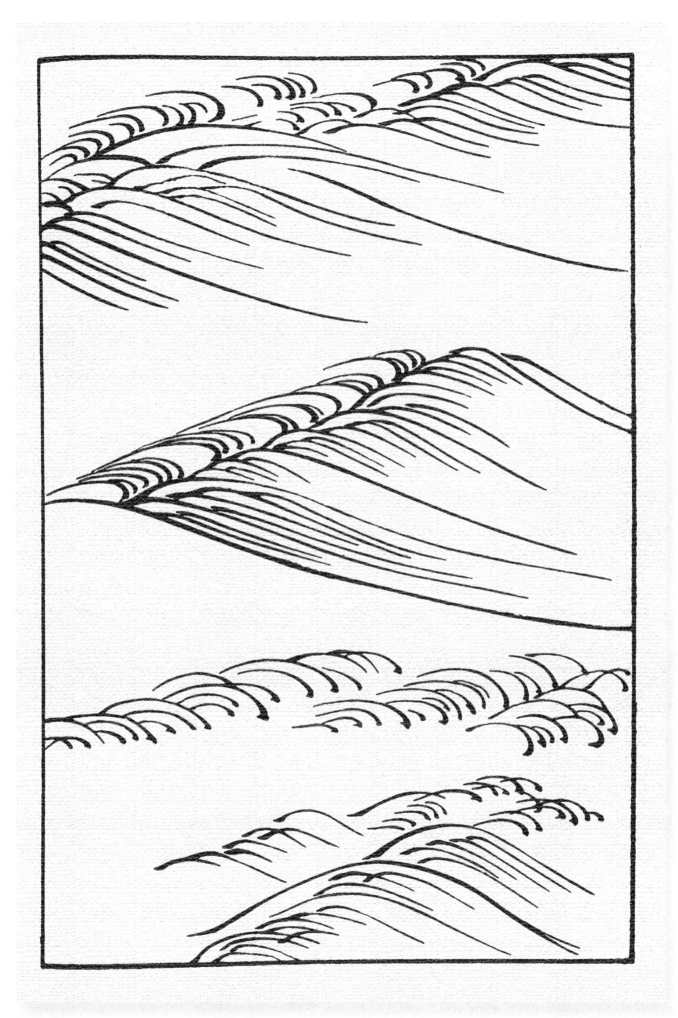

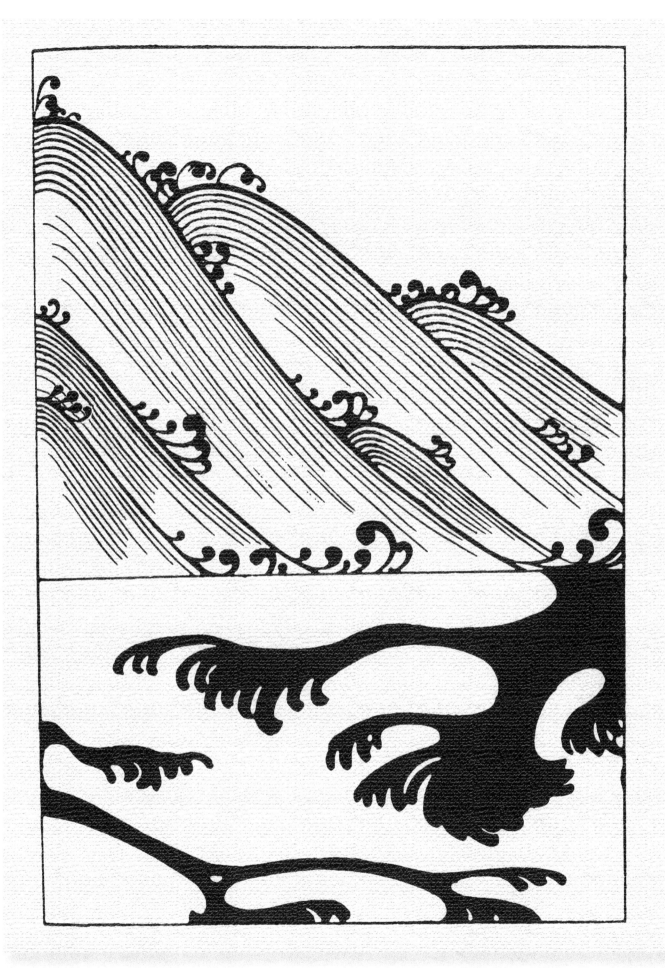

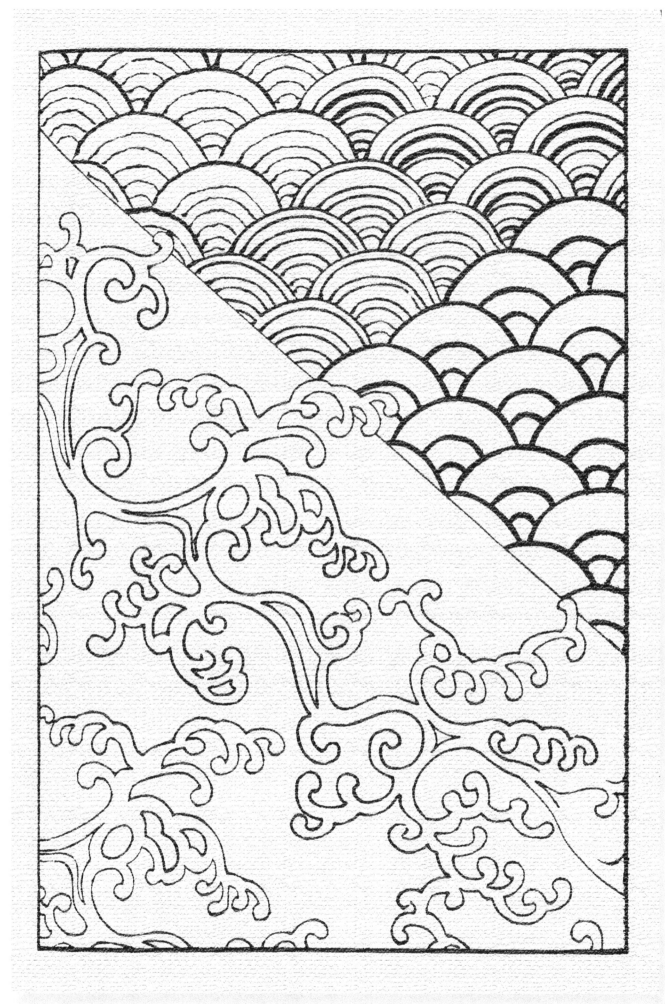

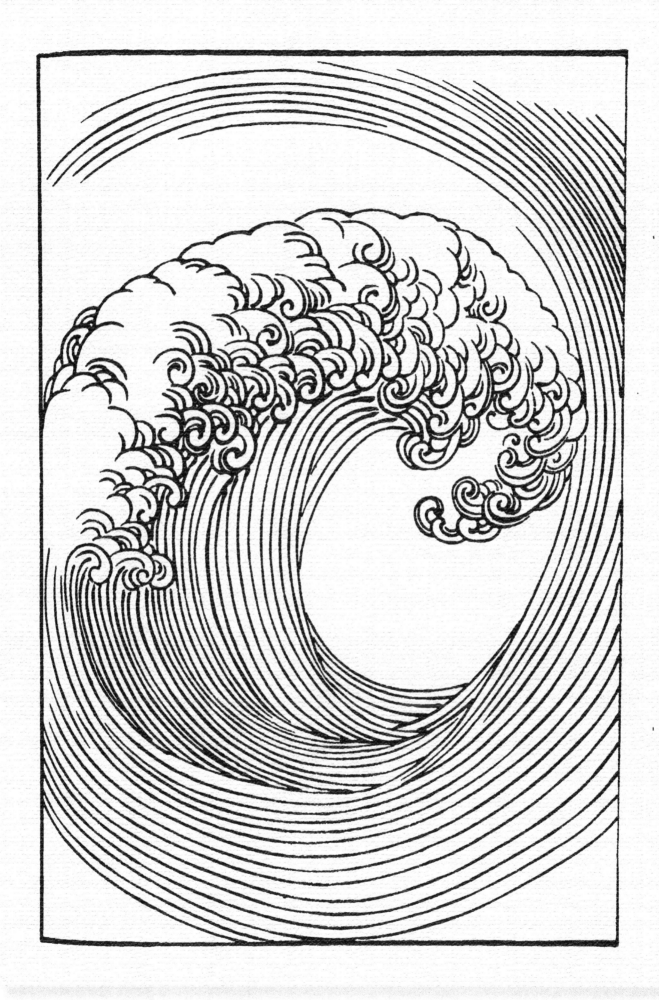

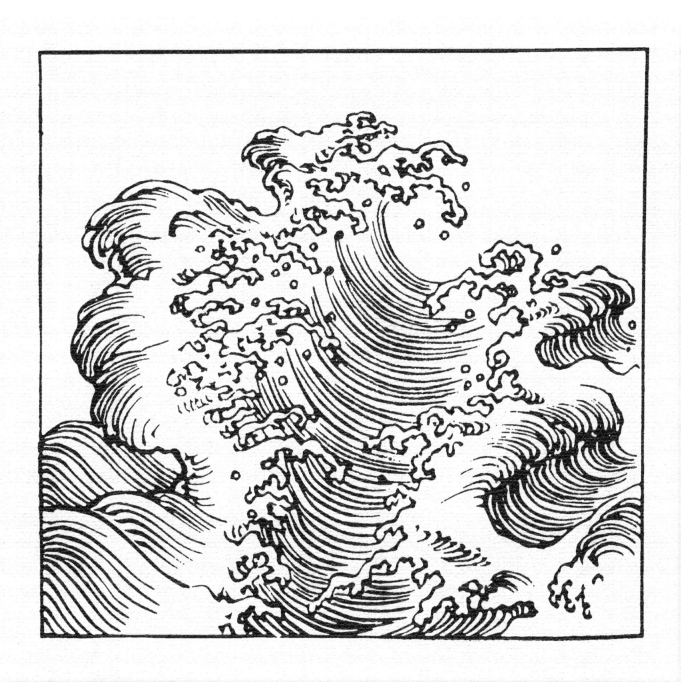

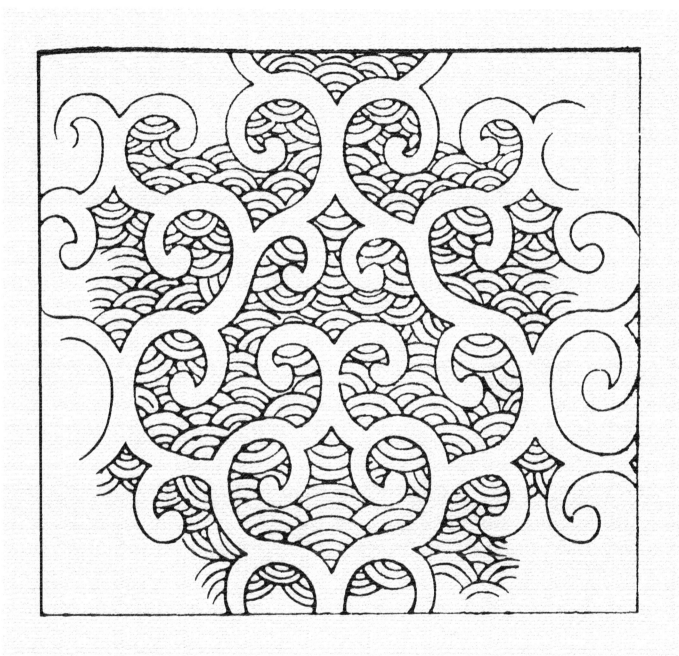

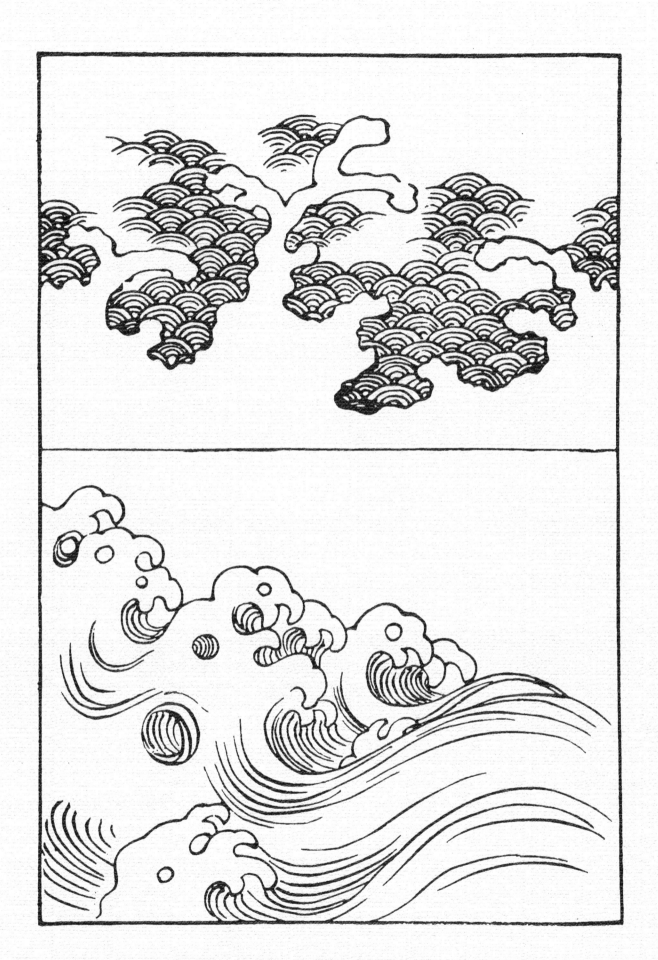

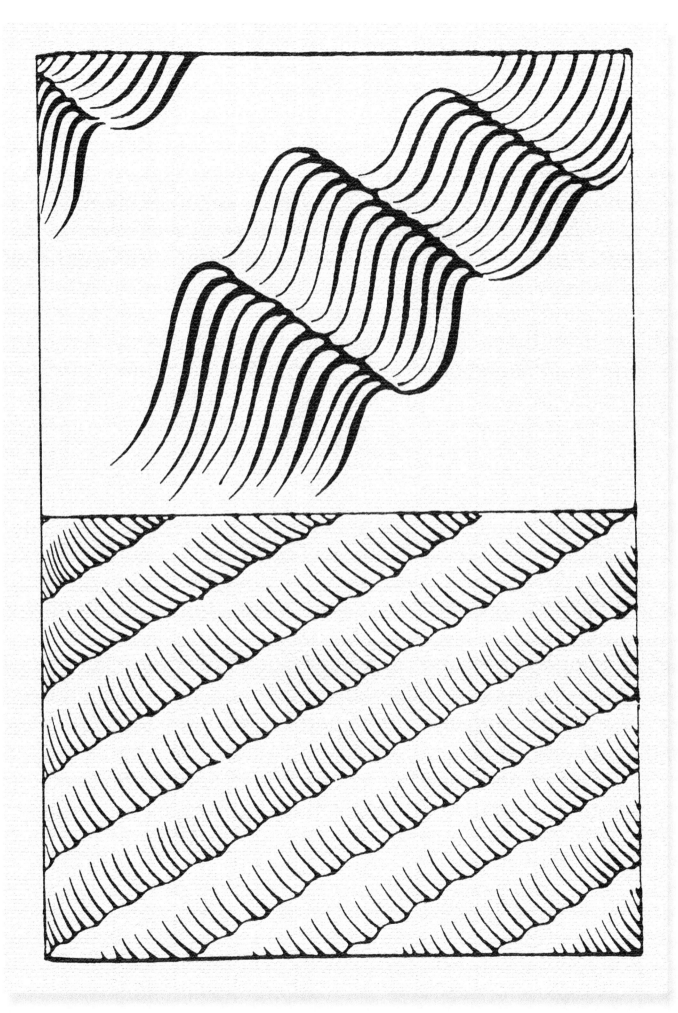

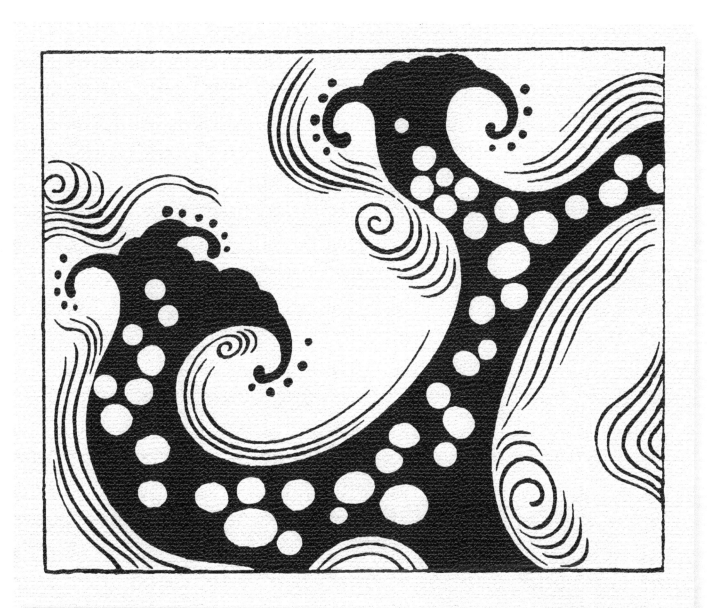

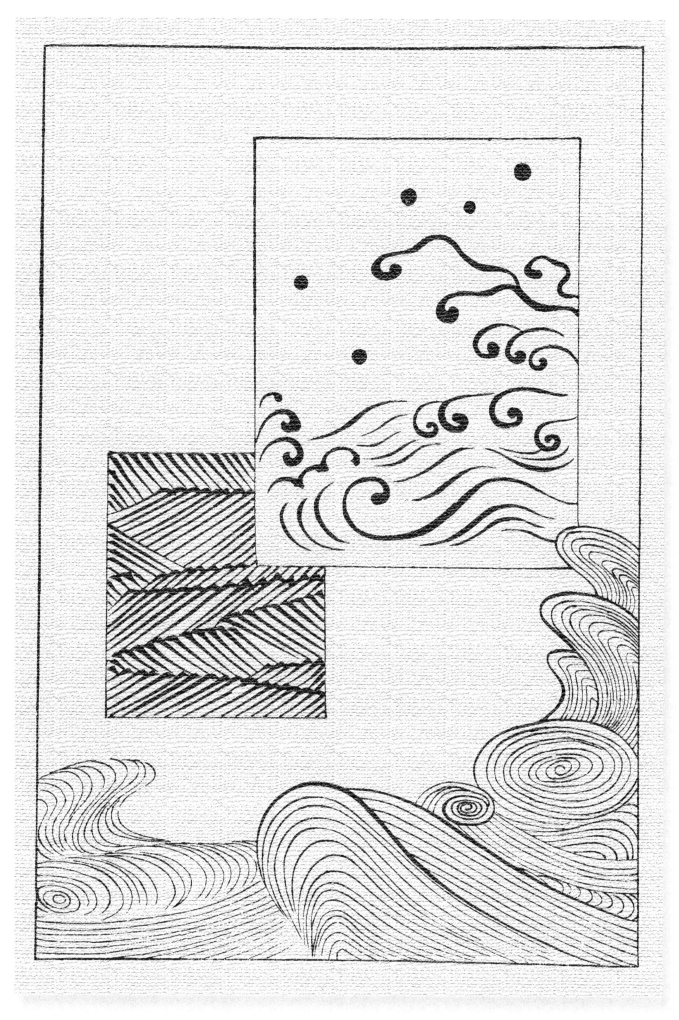

*five cured kippers dance*

*wild stallion hooves ripple*

*only one true north*

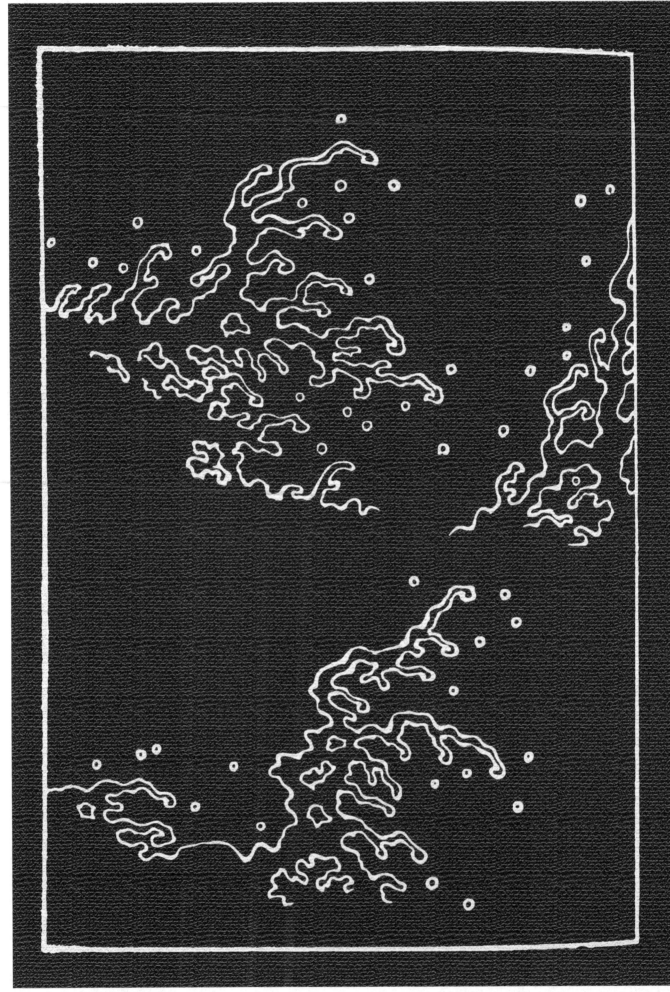

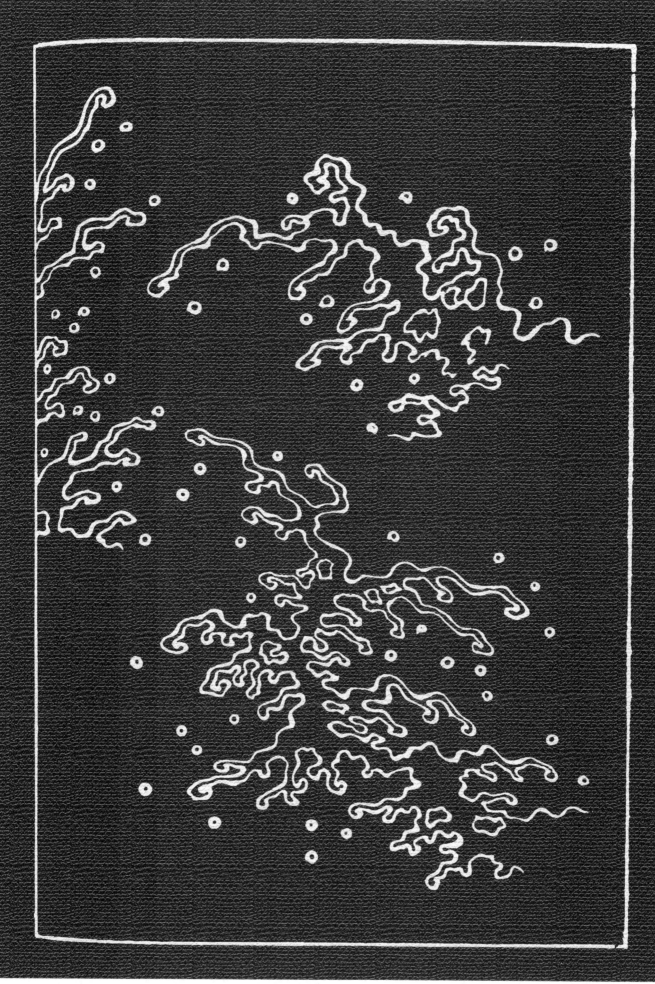

*treasure within sight*

*but to wealth a great distance*

*set waves in motion*

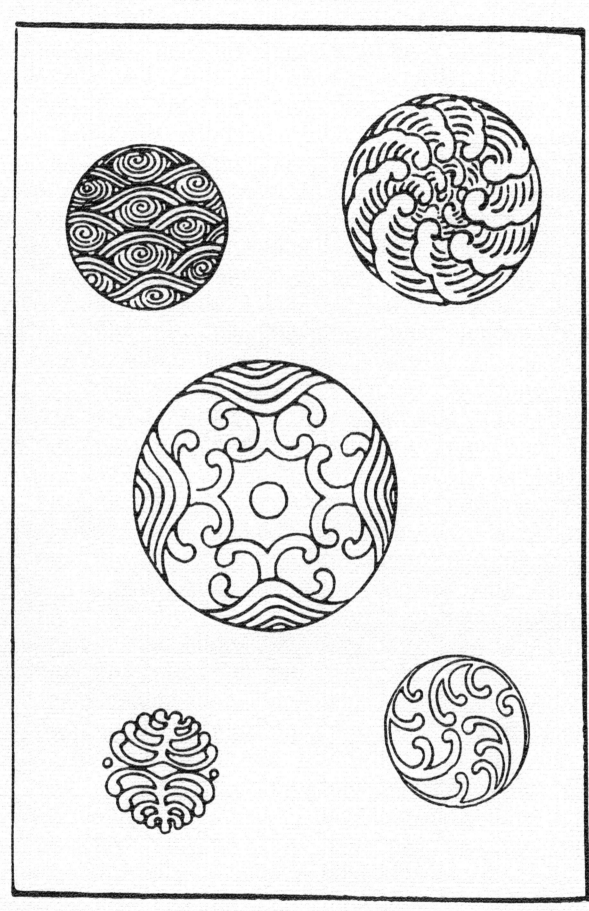

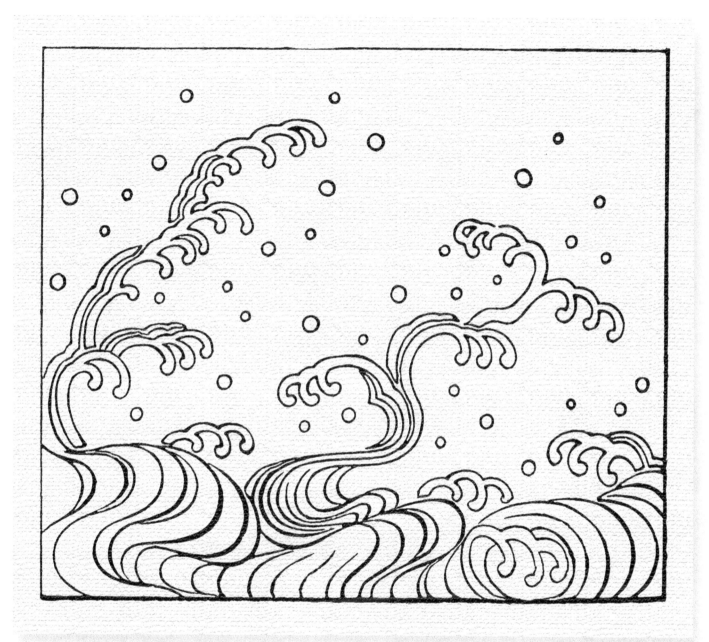

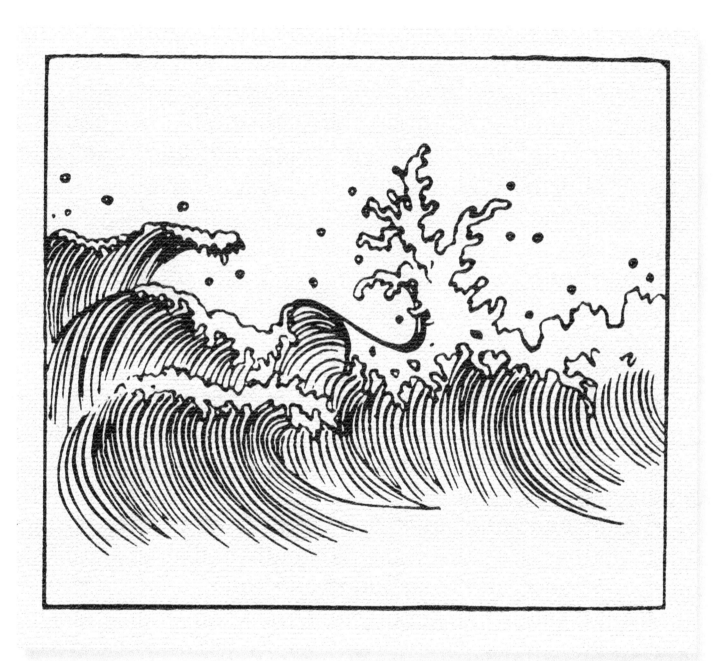

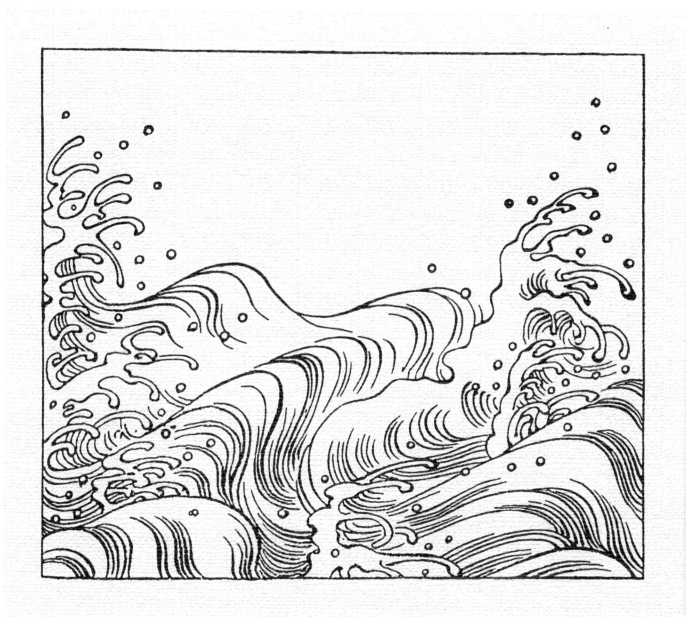

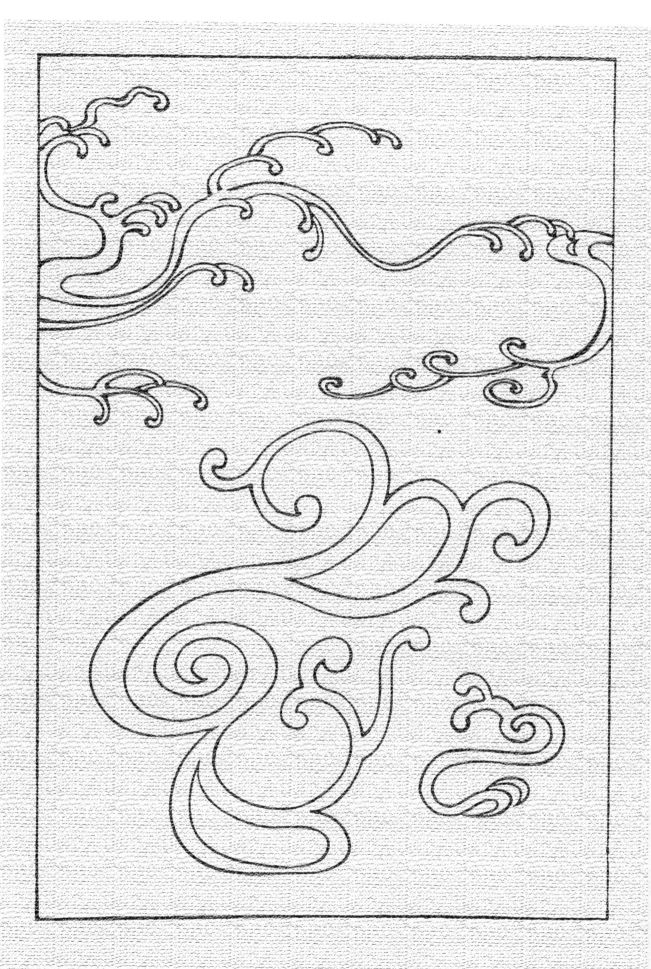

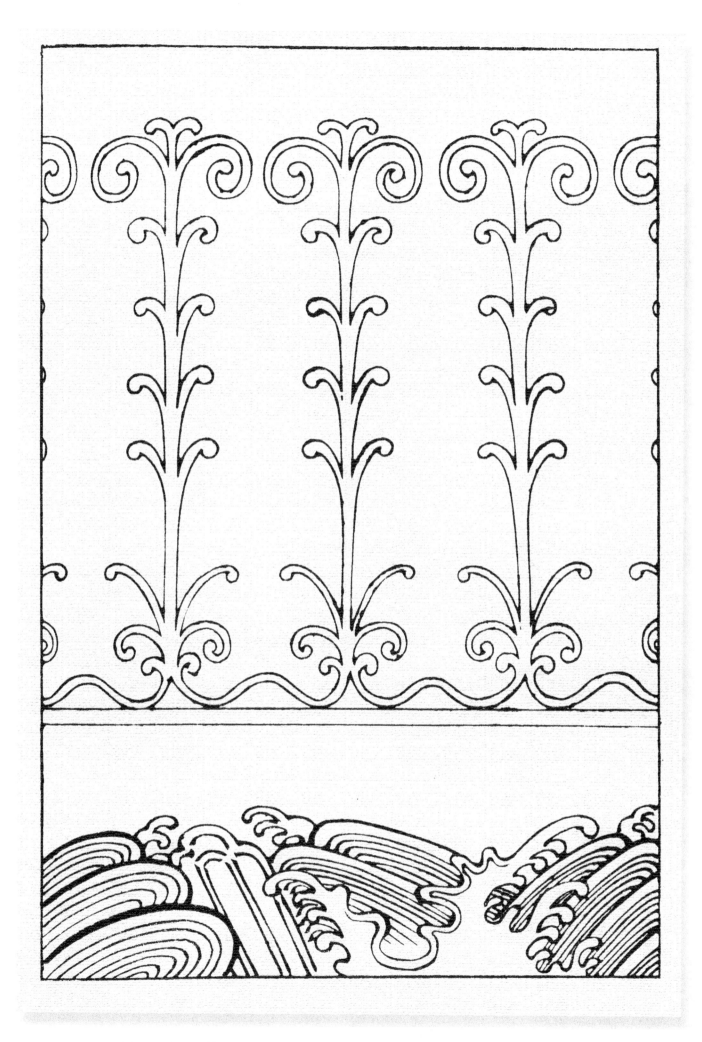

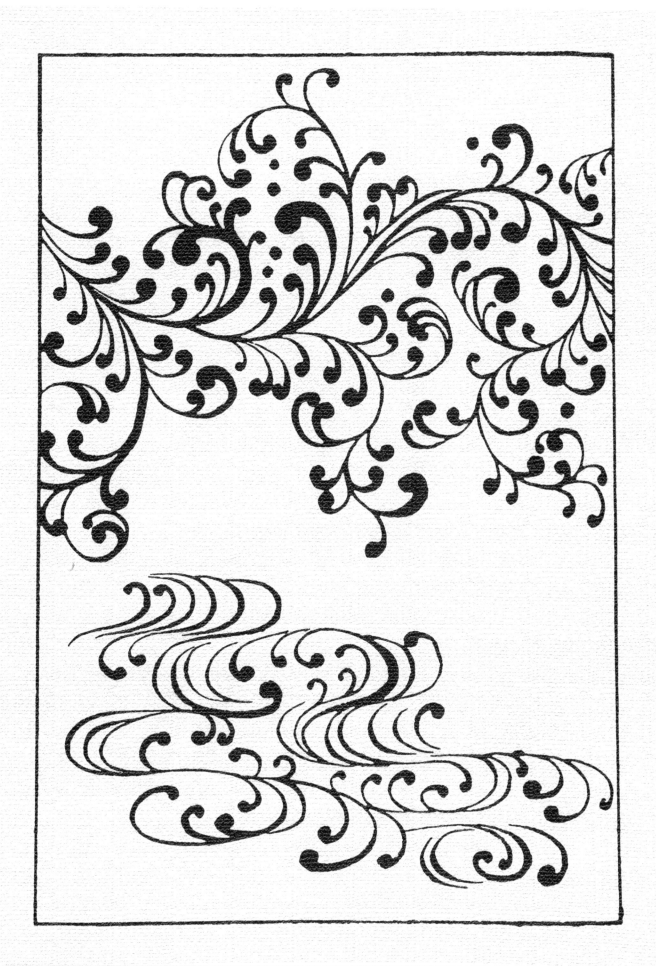

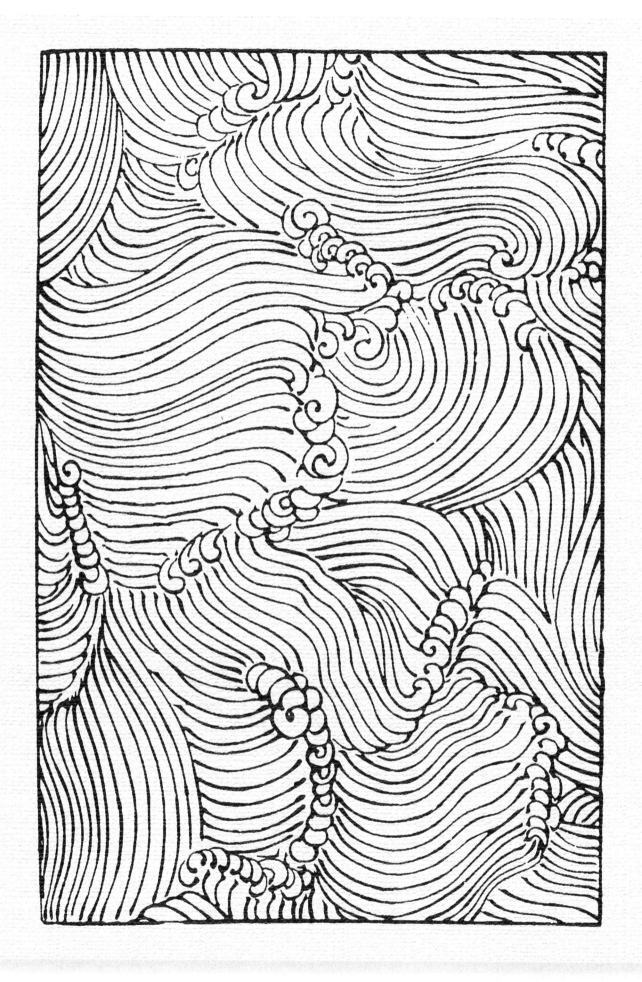

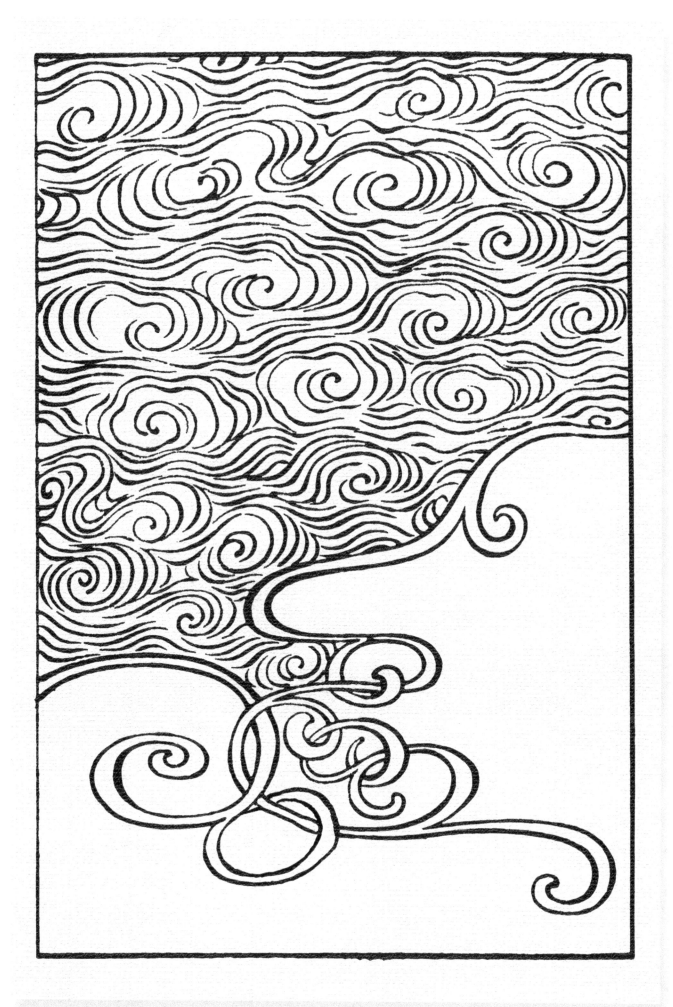

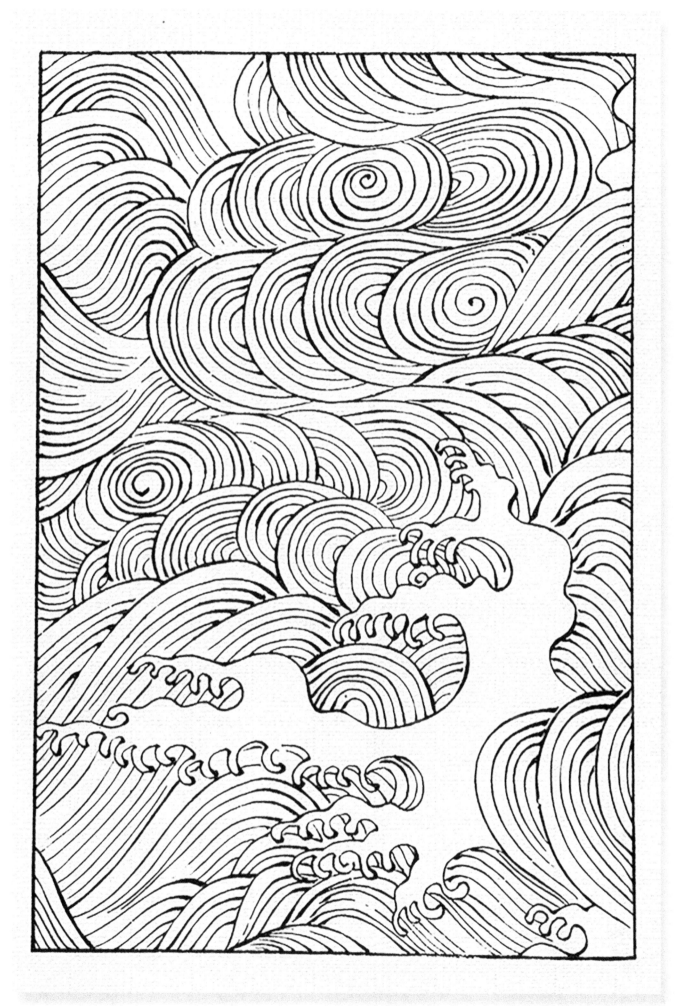

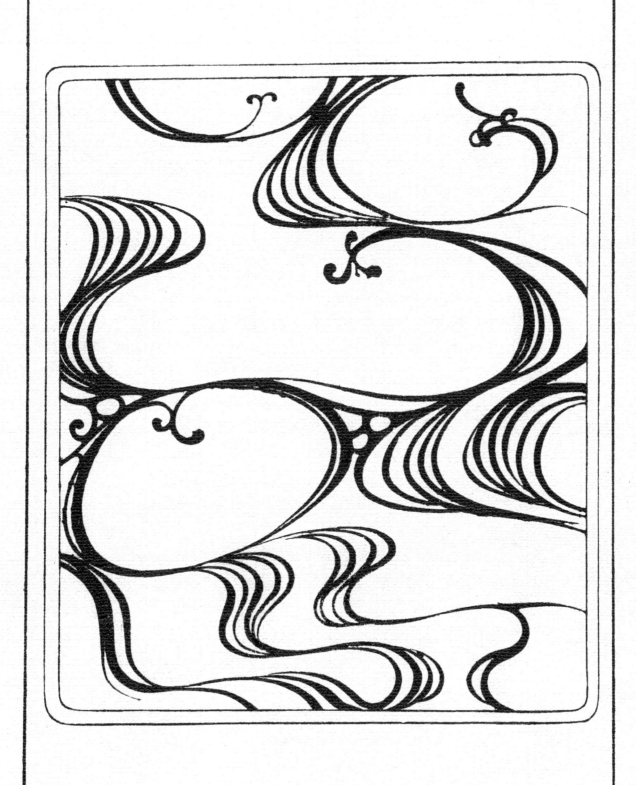

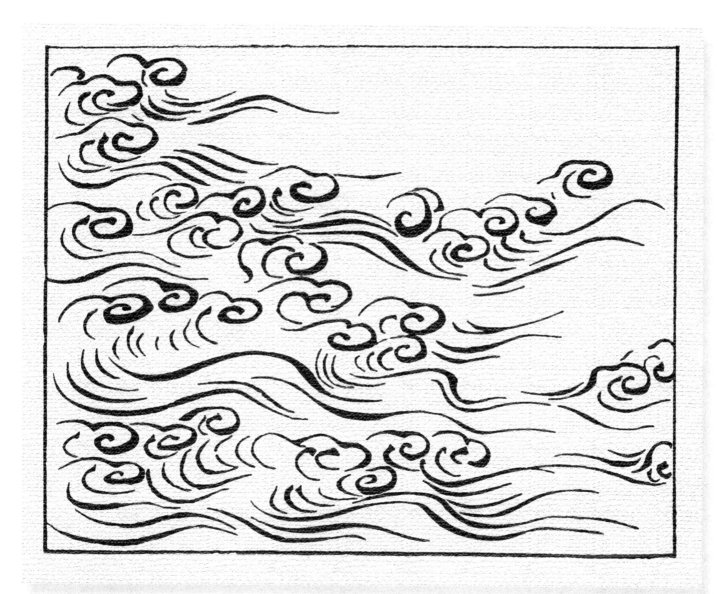

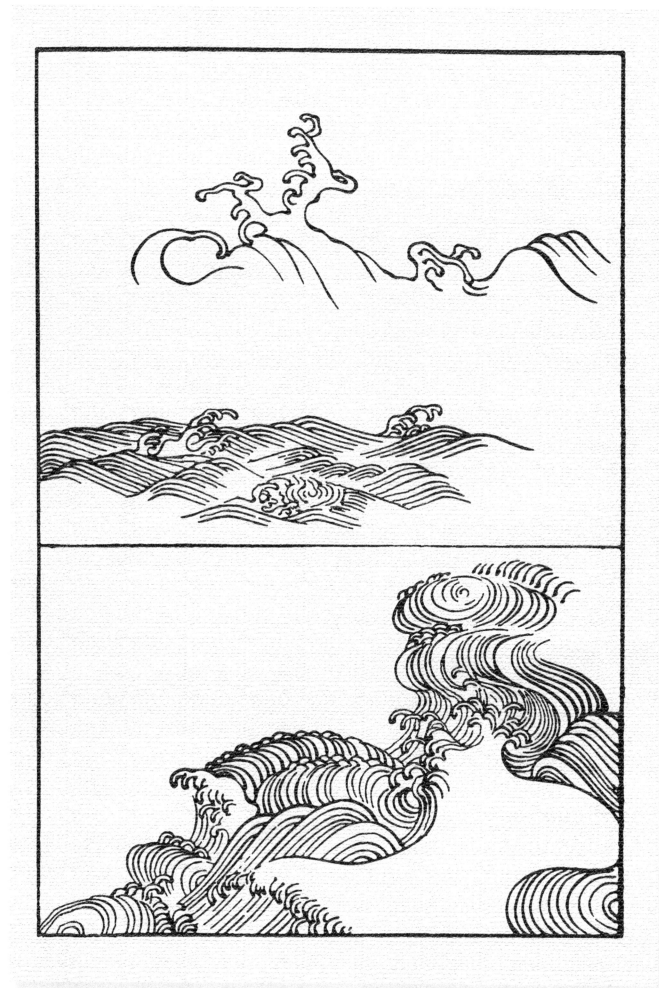

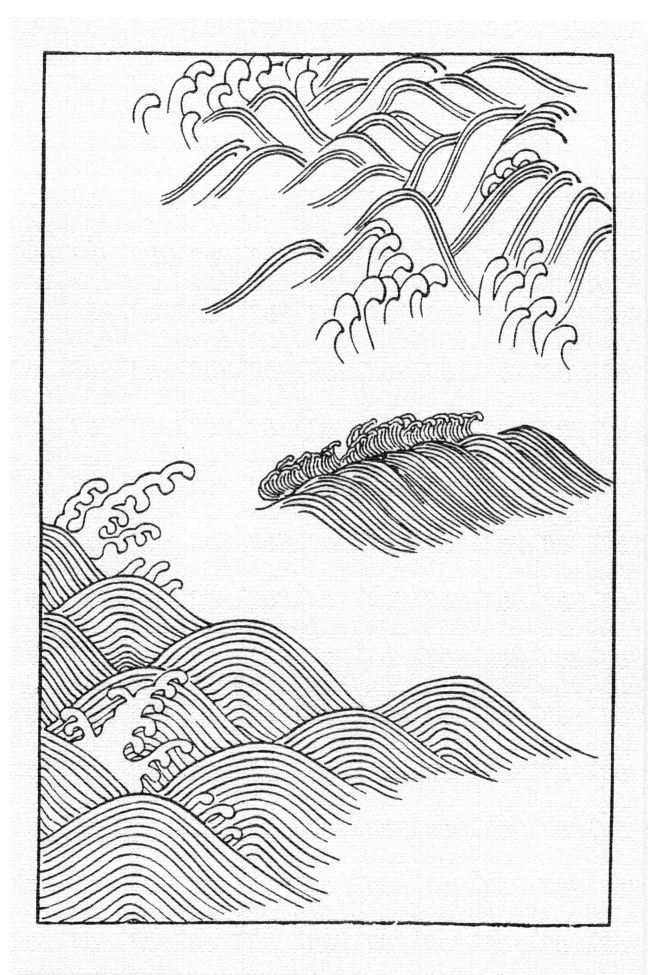

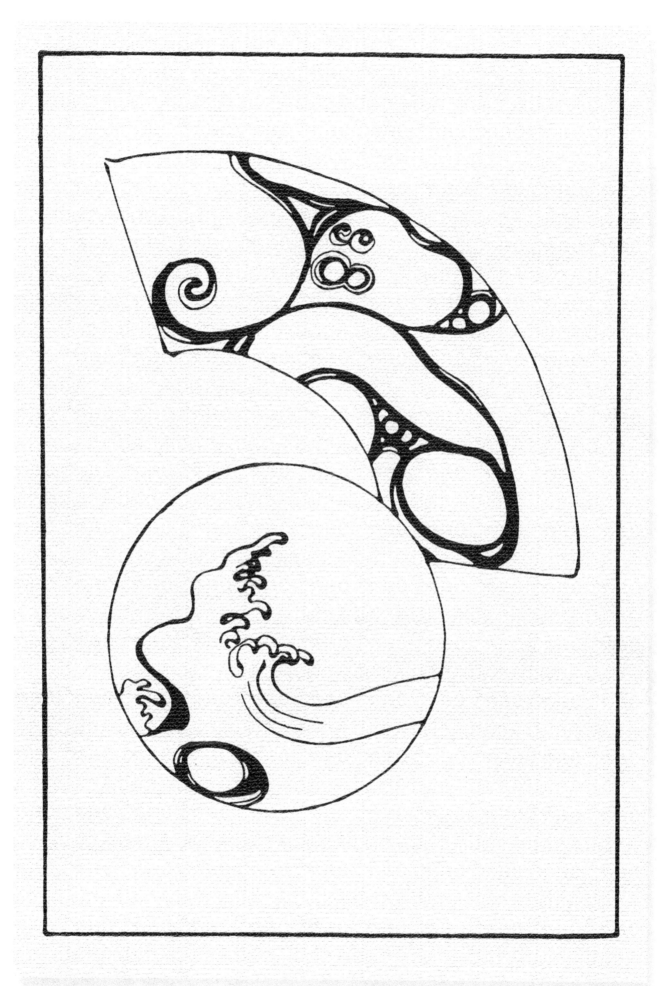

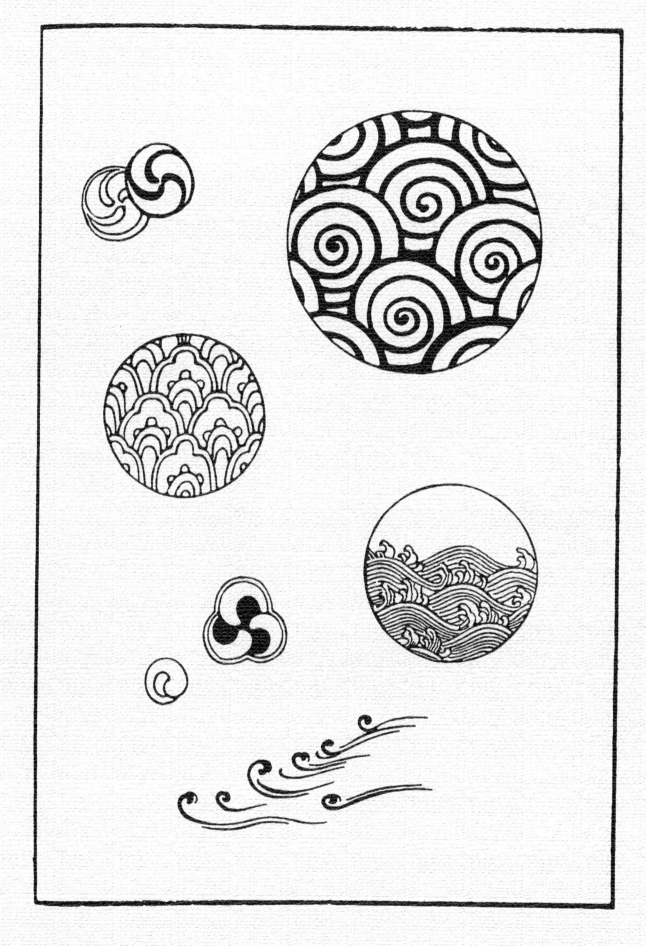

# RECOMMENDED BOOKS

*The Feyman Lectures on Physics, Volume 3* by Richard P. Feyman
*A Brief History Of Time* by Stephen Hawking
*The Odyssey* by Homer, Emily Wilson translation
*Kafka on the Shore* by Haruki Murakami
*The Sea Around Us* by Rachel Carson
*The Voyage of the Beagle* by Charles Darwin
*Log from the Sea of Cortez* by John Steinbeck
*20,000 Leagues Under the Sea* by Jules Verne
*The Shadow Line* by Joseph Conrad
*The Old Man and the Sea* by Ernest Hemingway
*The Perfect Storm* by Sebastian Junger
*The Ocean World* by Jacques Cousteau
*Frost & Fire* by Roger Zelazny
*In the Heart of the Sea: The Tragedy of the Whaleship Essex*
by Nathaniel Phibrick

# ABOUT THE AUTHOR

Satoshi Nakamoto is the renowned inventor of Bitcoin. He authored the original white paper titled, "Bitcoin: A Peer-to-Peer Electronic Cash System," implemented the first blockchain, and deployed Bitcoin in 2009 as the first decentralized digital currency. He is donating 100% of his book royalties to support STEM and environmental education programs serving underprivileged youth.

CPSIA information can be obtained
at www.ICGtesting.com
Printed in the USA
LVHW051407170619
621457LV00001BB/1/P

9 781945 652035